Abdelhafid Mimouni

When Bioinorganics Demystifies Alzheimer's

Abdelhafid Mimouni

When Bioinorganics Demystifies Alzheimer's

ScienciaScripts

Imprint

Any brand names and product names mentioned in this book are subject to trademark, brand or patent protection and are trademarks or registered trademarks of their respective holders. The use of brand names, product names, common names, trade names, product descriptions etc. even without a particular marking in this work is in no way to be construed to mean that such names may be regarded as unrestricted in respect of trademark and brand protection legislation and could thus be used by anyone.

Cover image: www.ingimage.com

This book is a translation from the original published under ISBN 978-620-6-72110-9.

Publisher:
Sciencia Scripts
is a trademark of
Dodo Books Indian Ocean Ltd. and OmniScriptum S.R.L publishing group

120 High Road, East Finchley, London, N2 9ED, United Kingdom
Str. Armeneasca 28/1, office 1, Chisinau MD-2012, Republic of Moldova, Europe
Printed at: see last page
ISBN: 978-620-8-09953-4

Bioinorganics demystifies Alzheimer's disease

Author : Dr. Abdelhafid Mimouni

An independent researcher in bioinorganic chemistry, Dr Mimouni is an expert in macromolecular synthesis and characterisation. He obtained his PhD in Chemistry from the University of Paris XII in 1997, after a Diplôme des Études Approfondies in Bioinorganic Systems from the University of Paris XI in 1993, where he also obtained his Licence and Maîtrise in Chemistry.

Book summary

This book explores Alzheimer's disease through the prism of bioinorganic chemistry, examining its history, global impact and current challenges. It details the biology of amyloid plaques and tau neurofibrils, as well as the role of metals such as copper in neurodegenerative processes. The focus is on copper, discussing its involvement in amyloid plaques, its disrupted homeostasis and its effects on neurons. The book analyses metal deposits in the brain using X-ray absorption spectroscopy (XAS). In conclusion, it summarises current knowledge, the implications for research and treatment, and suggests avenues for future research. Although I am an expert in XAS, my involvement is limited because I am not involved in research using synchrotron radiation or particle accelerators, which restricts my role in these new directions.

Map of the book

Chapter 1: Introduction to Alzheimer's disease

History of Alzheimer's disease

Alzheimer's disease, first identified in 1906 by the German neurologist Alois Alzheimer, has become one of the most studied neurodegenerative diseases of the 21st century. During the autopsy of a deceased patient, Alzheimer observed amyloid plaques and neurofibrillary tangles in the brain, features now recognised as pathological markers of the disease. Since this discovery, our understanding of the disease has evolved considerably, with major advances in biochemistry, neurobiology and, more recently, bioinorganics. However, despite more than a century of research, the precise mechanisms behind the disease remain largely unknown, and no cure has yet been found.

Worldwide statistics

Alzheimer's disease is now a global health crisis. In 2023, it was estimated that over 55 million people worldwide were living with the disease, a figure that could rise to 139 million by 2050. In France, around 1.2 million people have the disease, with prevalence rising steadily as the population ages. Alzheimer's disease represents an immense burden for healthcare systems and families, with considerable economic and social costs. People aged over 65 are the most affected, but early forms, although rare, can occur in younger individuals, often with a strong genetic component.

Contemporary issues

Longer life expectancy, the result of medical progress and improved living conditions, has led to a proportional increase in cases of dementia, of which Alzheimer's disease accounts for around 60-70%. This growth poses new challenges for healthcare systems, which must adapt to manage an ageing population. What's more, the complexity of the disease makes it difficult to develop effective treatments, despite the massive efforts of pharmaceutical research. The social stigma attached to people with dementia, and the emotional and financial burden placed on family carers, are also crucial issues that require a global response.

Research prospects

In this context, the bioinorganic approach is emerging as a promising avenue of research, likely to offer novel solutions. Bioinorganics, which explores the interaction between metallic elements and biological systems, could shed light on new aspects of neurodegenerative processes. Metals such as iron, copper and zinc play complex roles in the brain, both essential and potentially toxic. Imbalances in these metals could contribute to the pathologies observed in Alzheimer's disease, paving the way for innovative therapeutic strategies. Advanced techniques, such as X-ray absorption spectroscopy (XAS), now make it possible to examine these interactions with unprecedented precision, offering a new dimension to the understanding and potential treatment of this devastating disease. Exploring these avenues could not only improve patients' quality of life, but also slow or even halt the progression of the disease.

Chapter 2: Biochemical basis of Alzheimer's disease

Alzheimer's disease (AD) is a complex neurodegenerative disorder manifested principally by two types of brain lesions: amyloid plaques and tau-associated neurofibrillary degeneration. Although these biochemical abnormalities are distinct, they are deeply interconnected in the process of neurodegeneration.

Biology of Amyloid Plaques and Tau Protein Neurofibrils

Amyloid plaques are made up of fragments of beta-amyloid (Aβ) protein, resulting from the abnormal cleavage of amyloid precursor protein (APP). This cleavage is generally carried out by the enzymes β-secretase and γ-secretase, but deregulation of these processes can lead to the accumulation of insoluble Aβ fragments that aggregate in plaques between neurons. These plaques disrupt neuronal communication, lead to excessive activation of glial cells and trigger deleterious inflammatory responses. The accumulation of amyloid plaques is thus associated with a progressive decline in cognitive function and neuronal death.

Neurofibrils are abnormal structures found inside neurons and are mainly composed of the hyperphosphorylated protein tau. The tau protein, normally involved in stabilising microtubules, becomes hyperphosphorylated in Alzheimer's disease, leading to its dissociation from microtubules and their disorganisation. This disruption affects the transport of nutrients and substances essential for the proper functioning of neurons, contributing to neuronal degeneration and the formation of neurofibrillary degeneration. This phenomenon is particularly visible in the advanced stages of the disease.

The Role of Metals in the Brain: Iron, Copper, Zinc, and their Implication in Neurodegenerative Processes

Essential metals such as **iron**, **copper** and **zinc** play a crucial role in brain biochemistry. Iron is involved in fundamental biological processes, including cellular respiration and the synthesis of neurotransmitters. However, excessive accumulation of iron in the brain can catalyse the formation of free radicals via Fenton reactions, thereby increasing oxidative stress and causing oxidative damage to brain cells.

Copper is also essential, participating in various enzymatic functions, notably in antioxidant processes. However, copper overload is observed in amyloid plaques, where it stabilises beta-amyloid aggregates, exacerbating neurotoxicity. Zinc, although important for enzyme function and neurotransmission, can also accumulate in amyloid plaques, contributing to plaque stability and toxicity.

These metal imbalances are associated with neurodegenerative processes such as increased oxidative stress, disruption of neuronal function and the formation of amyloid plaques and neurofibrillary degeneration.

Biochemical Mechanisms: Complexity and the Current State of Knowledge

The biochemical mechanisms underlying Alzheimer's disease are extremely complex and multidimensional. Interactions between amyloid plaques, tau neurofibrils, metals and other biomolecules form an intricate network of biochemical reactions. These processes include :

- **Accumulation of misfolded proteins** : Misfolded beta-amyloid and tau proteins interact to form toxic aggregates.
- **Inflammation**: Amyloid plaques induce an inflammatory response which contributes to neuronal degeneration.
- **Mitochondrial Dysfunction**: The mitochondria, responsible for energy production, suffer damage, reducing cellular function and contributing to oxidative stress.
- **Disruption of Cell Signalling Pathways**: Critical signalling pathways are affected, leading to disturbances in cell regulation and neuronal death.

Despite the progress made in understanding these mechanisms, there are still some grey areas, particularly as regards the initial onset of the disease and the exact relationship between the various biochemical markers.

The Limits of Current Treatments: A Critical Review of Therapies Based on the Amyloid Hypothesis

Current therapies for Alzheimer's disease are mainly based on the amyloid hypothesis, which assumes that reducing beta-amyloid plaques will slow the progression of the disease. However, clinical results have been largely disappointing. The drugs available provide only temporary symptomatic relief and do not prevent the progression of the disease.

This critical assessment highlights the need to explore new therapeutic avenues, in particular those based on :

- **Moderation of metals in the brain**: By targeting the abnormal accumulation of metals such as copper and zinc.
- **Oxidative Stress Reduction**: Using antioxidants to limit oxidative damage.
- **Prevention of Tau hyperphosphorylation**: By developing treatments to modulate phosphorylation of the tau protein.

Chapter 3: The role of copper in Alzheimer's disease

Copper, an essential metal for the body to function properly, plays a crucial role in numerous enzymatic reactions, particularly in the brain. However, its disturbed homeostasis is associated with neurodegenerative pathologies such as Alzheimer's disease. This chapter explores in depth the involvement of copper in the formation of amyloid plaques, oxidative stress and its neurotoxic effects, as well as the potential implications for the treatment of the disease.

Copper in Amyloid Plaques

Copper is a cofactor for several enzymes, but in the context of Alzheimer's disease, it plays an ambivalent role. Studies have shown that copper binds to beta-amyloid peptides, stabilising amyloid plaques and possibly accelerating their formation. In the presence of copper, these plaques can also catalyse the production of free radicals via Fenton-like reactions, exacerbating oxidative stress and causing neuronal damage.

Copper's ability to interact with beta-amyloid leads not only to more rapid aggregation of these peptides, but also to the creation of toxic forms of beta-amyloid, which are more difficult for cellular clearance systems to degrade. As a result, copper promotes a neurotoxic environment in the brains of Alzheimer's patients, aggravating symptoms and accelerating the progression of the disease.

Disrupted homeostasis and oxidative stress

Copper homeostasis, the delicate balance between its absorption, storage and excretion, is disturbed in Alzheimer's disease. An excessive accumulation of copper, or an abnormal distribution of this metal in the brain, leads to an increase in oxidative stress. Excess copper can generate free radicals that damage membrane lipids, proteins and neuronal DNA.

Copper-induced oxidative stress is particularly harmful to neurons because of their sensitivity to oxidative damage. Once damaged, neurons enter a cycle of degeneration that results in a loss of cognitive function. Consequently, the accumulation of copper in specific regions of the brain is strongly correlated with the intensity of clinical symptoms observed in Alzheimer's patients.

Comparison with other metals: Iron and Zinc

The role of copper in Alzheimer's disease cannot be fully understood without comparing it with other essential metals, such as iron and zinc. Iron, like copper, is involved in redox reactions that can generate free radicals. However, iron is mainly involved in the respiratory function of mitochondria and its abnormal accumulation is often associated with mitochondrial dysfunction in Alzheimer's disease. Zinc also binds to beta-amyloid, but its role is less clear. Some researchers suggest that zinc may have a protective effect by inhibiting the activity of copper in catalysing free radicals.

Despite these similarities, copper stands out for its higher redox potential and its ability to participate in harmful biochemical reactions in the brains of Alzheimer's patients. So, although iron and zinc are also involved in pathological processes, copper appears to play a more direct and potentially more harmful role.

Therapeutic implications

Understanding the role of copper in Alzheimer's disease opens the door to new therapeutic strategies. Managing copper homeostasis could be a promising approach to slowing or even preventing the progression of the disease. Copper chelating agents capable of binding specifically to free copper in the brain are currently under investigation. These agents could potentially reduce oxidative stress by eliminating excess copper and preventing its involvement in the formation of amyloid plaques.

In addition, modulating copper absorption via targeted diets or supplements could offer a complementary approach to treating the disease. However, it is essential to maintain a balance, as copper deficiency could also have deleterious effects on neuronal health.

Chapter 4: Copper-based therapeutic approaches

Integrated and expanded text: Strategies for restoring copper homeostasis, with a focus on chelators

Disturbed copper homeostasis in the brain is a prominent feature of Alzheimer's disease. Copper, which is essential for many biological processes, becomes toxic when it is poorly regulated, contributing to the accumulation of amyloid plaques and exacerbating oxidative stress. One of the most promising approaches to restoring copper balance is the use of chelating agents. These compounds have the ability to bind to metal ions, thereby reducing their free concentration and minimising their harmful effects.

Copper chelators have attracted growing interest as a potential therapeutic approach for Alzheimer's disease. For example, the bis(thiosemicarbazone) complex (Cu(II)ATSM), initially developed to target oxidative stress in models of neurodegenerative disease, has shown promising results in preclinical studies. As well as modulating copper levels, this chelator also appears to influence other pathological processes, such as inflammation and mitochondrial dysfunction.

However, despite the significant progress made in the development of chelators, a number of challenges remain. The specificity of chelators is crucial to avoid undesirable side effects, as copper plays a fundamental role in many enzymatic functions. In addition, the transport of these molecules across the blood-brain barrier is a major obstacle. It is therefore necessary to develop chelators capable of specifically targeting the brain without disturbing copper levels in other organs.

Case studies: Analysis of molecules tested and their results in the laboratory

Several case studies highlight the efforts being made to test copper chelators in animal models and in vitro. One of the most extensively studied molecules is clioquinol, an anti-infective agent reused for its potential to modulate copper and zinc levels in the brain. Preclinical trials have shown that clioquinol reduces the accumulation of amyloid plaques and improves cognitive function in mouse models of Alzheimer's disease. However, results from the clinical phase have been mixed, highlighting the complexities of translating preclinical discoveries to human patients.

Another notable example is the study of PBT2, a clioquinol derivative. This compound has demonstrated an ability to restore metal homeostasis in the brain while improving cognitive function. Although phase 2 clinical trials have shown promising results, progress towards advanced phases has been hampered by questions of long-term safety and efficacy.

Challenges and prospects: The obstacles to be overcome to transform these discoveries into clinical treatments

The transition of copper chelators from the laboratory to the clinic involves many challenges. One of the main obstacles is optimising the bioavailability and tissue distribution of chelators. The development of specific vectors capable of transporting these molecules across the blood-brain barrier without disrupting metal homeostasis in the rest of the body is a priority.

In addition, individual variability in response to chelators, depending on genetics, stage of disease and other factors, complicates the development of universal treatments. Personalised approaches, tailored to patients' specific needs, could offer a more effective solution.

Finally, the integration of chelators into combined therapies, combined with other neuroprotective or anti-inflammatory agents, represents a promising way of maximising the efficacy of treatments while minimising side effects. Future research should also explore the long-term effects of copper modulation on neurodegeneration to ensure that these approaches do not disrupt other essential neurological functions.

Chapter 5: Using XAS to Analyse Metal Deposits in the Brain

Introduction to XAS: Basic principles of X-ray absorption spectroscopy

X-ray absorption spectroscopy (XAS) is a powerful technique for studying the presence and chemical specificity of metals in various biological environments. The technique is based on the absorption of X-rays by a sample, followed by analysis of the absorbed energy to determine the composition and local structure around specific atoms, such as copper, iron or zinc. XAS is divided into two main sub-techniques: near edge absorption spectroscopy (XANES) and extended absorption spectroscopy (EXAFS).

XANES is particularly useful for determining the oxidation state and chemical coordination of metals, while EXAFS provides an understanding of the local environment of metal atoms, including interatomic distances and bond types. This information is crucial for a better understanding of how metals, such as copper, are involved in pathological processes such as the formation of amyloid plaques in Alzheimer's disease.

Analysis of copper deposits in the brain: How XAS can be used to detect and quantify these deposits

In the context of Alzheimer's disease, XAS is proving to be an indispensable tool for analysing copper deposits in the brain. It has been shown that the brains of Alzheimer's patients contain up to five times more copper than healthy brains. XANES enables the concentration of copper to be precisely monitored, revealing not only the total quantity of metal, but also its oxidation state and its interaction with other biomolecules.

In addition, EXAFS provides a detailed view of the chemical environment of copper, identifying neighbouring atoms and bonding distances. This analysis is crucial for understanding how copper becomes incorporated into amyloid plaques, potentially exacerbating their toxicity and contributing to oxidative stress.

Copper determination by XANES: Methodology and Application

Introduction to XANES Spectroscopy

X-ray absorption spectroscopy (XAS) is a powerful technique for analysing the concentration of metals in various biological samples. The X-ray Absorption Near Edge Structure (XANES) region of the XAS spectrum provides valuable information about the concentration of metals and their local environment. For the determination of copper in brain deposits, recording the XANES spectrum provides data specific to the presence and concentration of copper in the samples.

Analysis Methodology

Recording the XANES spectrum :

To measure the concentration of copper in mouse brain aggregates, it is essential to capture the XANES spectrum of samples containing copper deposits. The XANES spectrum is obtained by exposing the sample to X-rays and measuring absorption at energies close to the absorption threshold of copper.

Measuring Threshold Intensity :

The intensity of the threshold in the XANES spectrum is used to assess the concentration of copper in the sample. According to Beer-Lambert's law, the absorption of X-rays by the sample is proportional to the concentration of the target metal. By measuring this intensity and applying Beer-Lambert's law, it is possible to quantify the concentration of copper in brain deposits.

Concentration-based measurement techniques

XAS Fluorescence Method :

When the concentration of copper in the sample is less than 10 mM, the

XAS fluorescence method is recommended. This technique is suitable for low concentrations and allows sensitive detection of copper by measuring the X-rays emitted by the sample after absorption of the incident X-rays.

Transmission method in XAS :

For samples with copper concentrations above 10 mM, the XAS transmission method is more appropriate. This approach involves the measurement of X-rays transmitted through the sample, providing information on the total copper uptake in the sample. This method is effective for higher concentrations and offers greater accuracy in assessing copper levels.

Conclusion

By using XANES to measure copper, it is possible to obtain quantitative data on the concentration of this metal in brain deposits. Depending on the copper concentration, fluorescence or transmission methods are chosen to optimise the measurement and guarantee accurate results. These techniques provide crucial information for understanding the role of copper in Alzheimer's disease and for developing therapeutic approaches based on the regulation of this metal.

Analysis of the Copper Absorption Threshold and Determination of the Coordination Geometry Using XANES

The absorption threshold for pure copper is 8,979 eV. To assess the copper environment in our samples, we performed an energy scan ranging from 8,800 to 10,000 eV. Once the XANES spectrum has been obtained, interpretation of the absorption threshold maximum is used to determine the coordination of the copper. If the spectrum shows a distinct maximum at the absorption threshold with a white area, this indicates that the copper is in a perfect octahedral environment. On the other hand, if the spectrum shows a double peak, this suggests that the octahedron is distorted due to the degeneracy of the ppp atomic orbitals being lifted. This deformation can provide crucial information about the structural modifications and local interactions of copper in the samples analysed.

Experimental studies: Examples of research using XAS in the context of Alzheimer's disease

Studies on mouse models of Alzheimer's disease have provided valuable information on the role of copper and its accumulation in the brain. For example, transgenic mice expressing high levels of amyloid proteins have been used to study the effects of copper chelation. XAS, and in particular XANES, was used to monitor changes in copper concentrations before and after treatment with chelators, demonstrating a significant reduction in copper levels in amyloid plaques after chelation.

Another study used EXAFS to examine the interactions between copper and amyloid proteins in mouse brains. The results showed that copper binds to specific sites within amyloid plaques, altering their conformation and increasing their propensity to generate reactive oxygen species (ROS). These studies highlight the importance of copper regulation in the brain and offer prospects for the development of therapies based on metal modulation.

Future research avenues: Proposals for new studies using XAS to better understand the distribution and impact of metals in the brain

In the future, XAS could be used to explore several as yet unresolved aspects of the biology of metals in Alzheimer's disease. For example, studies could focus on the spatial variations in copper concentration in different brain regions, correlated with disease progression. In addition, combining XAS with other imaging techniques, such as MRI or positron emission tomography (PET), could provide an overview of the dynamics of metals and their role in neurodegeneration.

It would also be interesting to explore the effect of new metal chelators on the chemical environment of copper in the brain. XANES and EXAFS could be used to assess the effectiveness of these chelators in reducing copper toxicity and restoring metal homeostasis. This research would contribute to the design of more targeted and potentially more effective therapies for the treatment of Alzheimer's disease.

Chapter 6: Conclusion and future prospects

Summary of current knowledge: What science currently knows about the role of metals in Alzheimer's disease

Alzheimer's disease is a complex neurodegenerative disorder, and over the decades research has led to a better understanding of the underlying mechanisms, including the role of metals such as copper, iron and zinc. Amyloid plaques and tau protein tangles, pathological markers of the disease, have been associated with a disturbance in metal homeostasis in the brain. Studies using advanced techniques such as X-ray absorption spectroscopy (XAS) have revealed that these metals are not simply by-products of the disease, but play an active role in the progression of neurodegeneration.

The accumulation of copper in the brains of Alzheimer's patients, and its interaction with amyloid proteins, promotes the formation of toxic structures and oxidative stress, thereby exacerbating neuronal loss. Although current therapies focus primarily on the amyloid hypothesis, they have shown significant limitations, notably in failing to significantly alter the course of the disease. This situation highlights the need for new therapeutic approaches, particularly those which address the regulation of metals in the brain.

Implications for research and treatment: How bioinorganic approaches can transform our approach to this disease

Bioinorganic approaches, which examine the role of metals within biological systems, offer a unique and promising perspective for research into Alzheimer's disease. These approaches provide a better understanding of how metals, and copper in particular, interact with proteins and other biomolecules to influence the pathology of the disease. The use of techniques such as XAS has already demonstrated its potential to identify specific metal targets in the brain, paving the way for more targeted therapeutic interventions.

The management of copper homeostasis through the use of chelators or other regulatory agents could represent a major advance in the treatment of Alzheimer's disease. By directly targeting the underlying metal mechanisms, these strategies could not only slow the progression of the disease, but also prevent the onset of symptoms at an earlier stage.

Research proposals: Suggestions for future research, based on the gaps identified in the previous chapters

Future research could focus on several areas to fill the current gaps in our understanding of Alzheimer's disease:

1. **Exploring multi-metal interactions**: While research has largely focused on individual metals such as copper or iron, it is essential to explore how these metals interact with each other in the context of disease. Comparative studies using XAS to analyse several

metals simultaneously could reveal previously unknown interactions.

2. **Development of new selective chelators**: There is a need to design chelators that not only reduce copper levels, but do so selectively, without disrupting other important biological functions. Preclinical studies in mouse models, followed by clinical trials, would be crucial to assess their efficacy and safety.

3. **Longitudinal monitoring of metal deposits**: It would be beneficial to monitor metal levels in the brains of Alzheimer's patients over a long period, in order to correlate these data with the progression of the disease and the effectiveness of treatments. The use of XAS in longitudinal studies could provide essential information for adjusting therapies over time.

4. **Analysis of the side-effects of metal therapies**: Given that modulating metal levels in the brain can have wide-ranging effects, in-depth research is needed to assess the risks and benefits of bioinorganic therapies, in order to minimise potential side-effects.

Vision for the future: The next steps in advancing the treatment of Alzheimer's disease

The future of Alzheimer's treatment could be radically transformed by advances in our understanding of the role of metals in neurodegeneration. The next steps must include integrating bioinorganic discoveries into therapeutic strategies and developing treatment protocols that specifically target metal disturbances.

The development of new technologies, such as next-generation metal chelators and advanced imaging techniques, will make it possible to monitor disease progression in real time and adjust treatments accordingly. In addition, interdisciplinary collaboration between chemists, biologists, neurologists and clinicians will be essential to transform fundamental discoveries into effective clinical interventions.

In conclusion, although many challenges remain to be overcome, the prospects offered by bioinorganic approaches are promising. They could well represent the next great advance in the fight against Alzheimer's disease, offering patients better options for managing and potentially preventing this devastating neurodegeneration.

Glossary :

1. **Alzheimer's disease**: A neurodegenerative disease characterised by the progressive loss of memory, cognitive abilities and behavioural functions. It is associated with deposits of amyloid plaques and neurofibrillary tangles in the brain.

2. **Amyloid plaques**: Abnormal accumulations of beta-amyloid peptides in the brain, considered to be a key marker of Alzheimer's disease. They form extracellular deposits that contribute to neurodegeneration.

3. **Neurofibrillary tangles**: intracellular aggregates of hyperphosphorylated tau proteins which disrupt neuronal function by altering the microtubule network and contribute to neuronal degeneration.

4. **Bioinorganics**: Branch of chemistry that studies the role of metallic elements in biological systems, focusing on their interactions, distribution and functions in living organisms.

5. **X-ray absorption spectroscopy (XAS)**: Analytical technique used to study the local structure of atoms in a material by analysing the absorption of X-rays.

6. **Dementia**: A general term for a severe decline in mental ability, severe enough to interfere with daily life. Alzheimer's disease is the most common form of dementia.

7. **Life expectancy**: Average expected lifespan for a person in a given population, often used as a public health indicator.

8. **Social stigma**: The phenomenon whereby a person or group is marginalised or devalued because of a health condition, appearance or behaviour perceived as different.

9. **Family carers**: Members of the family who provide unpaid care to a dependent person, often because of a chronic illness or disability.

10. **Neurobiology**: Study of nervous systems and their functions, including the neural processes underlying cognition, behaviour and neurological diseases.

11. **Metals**: Elements such as iron, copper and zinc, whose abnormal accumulation in the brain is associated with Alzheimer's disease.

12. **Hyperphosphorylation**: Biochemical process in which an excess of phosphate groups is added to a protein, often altering its normal function.

13. **Oxidative stress**: Imbalance between the production of free radicals and the body's ability to neutralise them, leading to cell damage and disruption of normal biological processes.

14. **Amyloid hypothesis**: Theory postulating that the accumulation of beta-amyloid plaques in the brain is the main cause of Alzheimer's disease.

15. **Copper (Cu)** : Transition metal essential for the functioning of many enzymes, but potentially neurotoxic if accumulated excessively in the brain.

16. **Homeostasis**: A state of dynamic equilibrium in which a biological system maintains its internal conditions constant despite external variations, including the regulation of metals in the body to avoid toxic or deficient concentrations.

17. **Free radicals**: Unstable molecules that can damage cells by oxidising cellular components such as lipids, proteins and DNA.

18. **Fenton-like reactions**: Chemical reactions in which copper or iron catalyses the production of free radicals from peroxides.

19. **Copper chelators**: Chemical substances that bind specifically to copper, facilitating its elimination from the body and reducing the risk of toxicity associated with excessive accumulation of this metal.

20. **Chelation**: Process by which a molecule (chelator) binds to a metal ion, forming a stable complex that can be eliminated from the body.

21. **Blood-brain barrier**: Protective structure separating the blood from the central nervous system, regulating the passage of substances between the blood and the brain.

22. **Clioquinol**: An anti-infective agent reused for its potential to modulate copper and zinc levels in the brain.

23.**PBT2**: Derived from clioquinol, studied for its potential to restore metal homeostasis in the brain and improve cognitive function.

24.**XANES (X-ray Absorption Near Edge Structure)** : X-ray absorption spectroscopy technique used to determine the oxidation state and chemical coordination of metals, providing information about the local structure around metal atoms.

25.**EXAFS (Extended X-ray Absorption Fine Structure)**: A complementary technique to XANES, used to analyse the local environment of metal atoms.

References :

1. Alzheimer, A. (1907). Über eine eigenartige Erkrankung der Hirnrinde. *Allgemeine Zeitschrift für Psychiatrie und psychisch-gerichtliche Medizin, 64*, 146-148.

2. Bush, A. I. (2003). The metallobiology of Alzheimer's disease. *Trends in Neurosciences, 26*(4), 207-214. https://doi.org/10.1016/S0166-2236(03)00067-5

3. Bush, A. I. (2013). The metal theory of Alzheimer's disease. *Journal of Alzheimer's Disease, 33*(Suppl 1), S277-S281. https://doi.org/10.3233/JAD-2012-129011

4. Bush, A. I., & Tanzi, R. E. (2008). Therapeutics for Alzheimer's disease based on the metal hypothesis. *Neurotherapeutics, 5*(3), 421-432. https://doi.org/10.1016/j.nurt.2008.05.001

5. Collingwood, J. F., & Davidson, M. R. (2014). The role of iron in neurodegenerative disorders: Insights from synchrotron X-ray studies. *Frontiers in Pharmacology, 5*, 191. https://doi.org/10.3389/fphar.2014.00191

6. Crouch, P. J., Barnham, K. J., & Bush, A. I. (2007). Therapeutic treatments for Alzheimer's disease based on metal bioinorganic chemistry. *Inorganic Biochemistry, 101*(5), 943-950. https://doi.org/10.1016/j.jinorgbio.2007.01.022

7. Crouch, P. J., Savva, M. S., Hung, L. W., Donnelly, P. S., Mot, A. I., Parker, S. J., Greenough, M. A., Volitakis, I., Adlard, P. A., Cherny, R. A., Masters, C. L., & Bush, A. I. (2011). The Alzheimer's therapeutic PBT2 promotes amyloid-β degradation and GSK3 phosphorylation via a metal chaperone activity. *Journal of Neurochemistry, 119*(1), 220-230. https://doi.org/10.1111/j.1471-4159.2011.07410.x

8. Gauthier, S., Rosa-Neto, P., Morais, J. A., & Webster, C. (2021). *World Alzheimer Report 2021: Journey through the diagnosis of dementia.* Alzheimer's Disease International. https://www.alzint.org/u/WorldAlzheimerReport2021.pdf

9. Greenough, M. A., & Camakaris, J. (2009). Metalloproteomics and the complex role of metals in neurodegenerative diseases. *Biochemical Society Transactions, 37*(6), 1260-1263. https://doi.org/10.1042/BST0371260

10. Hardy, J., & Higgins, G. (1992). Alzheimer's disease: The amyloid cascade hypothesis. *Science, 256*(5054), 184-185. https://doi.org/10.1126/science.1566067

11. Hardy, J., & Selkoe, D. J. (2002). The amyloid hypothesis of Alzheimer's disease: Progress and problems on the road to therapeutics. *Science, 297*(5580), 353-356. https://doi.org/10.1126/science.1072994

12.Hutchinson, R. W., & Hyman, B. T. (2006). Comparative chemistry of amyloid plaques. *Neuroscience Letters, 102*(1), 13-18. https://doi.org/10.1016/j.neulet.2006.05.054

13.Lovell, M. A., Robertson, J. D., Teesdale, W. J., Campbell, J. L., & Markesbery, W. R. (1998). Copper, iron and zinc in Alzheimer's disease senile plaques. *Journal of the Neurological Sciences, 158*(1), 47-52. https://doi.org/10.1016/S0022-510X(98)00092-6

14.Opazo, C., & Greenough, M. A. (2010). Copper and Alzheimer's disease: Emerging mechanisms. *Inorganica Chimica Acta, 363*(6), 1241-1247. https://doi.org/10.1016/j.ica.2009.11.018

15.Penner, M. R., & Lai, C. S. W. (2020). Neurobiology of Alzheimer's disease. In *The SAGE Encyclopedia of Human Development* (pp. 104-109). SAGE Publications.

16.Price, J. L., & Morris, J. C. (1999). Tangles and plaques in nondemented aging and "preclinical" Alzheimer's disease. *Annals of Neurology, 45*(3), 358-368. https://doi.org/10.1002/1531-8249(199903)45:3<358::AID-ANA12>3.0.CO;2-X

17.Quintana, C., Bellefqih, S., Laval, J. Y., Guerquin-Kern, J. L., Wu, T. D., Avila, J., & Ferrer, I. (2006). Study of the localization of iron, ferritin, and hemosiderin in senile plaques by energy-filtering transmission electron microscopy and electron energy-loss spectroscopy. *Journal of Structural Biology, 153*(1), 42-54. https://doi.org/10.1016/j.jsb.2005.11.006

18. Roberts, B. R., Ryan, T. M., & Bush, A. I. (2012). The role of zinc, copper and iron in Alzheimer's disease. *Journal of Alzheimer's Disease, 32*(4), 713-726. https://doi.org/10.3233/JAD-2012-120823

19. Ruloff, A., Stocker, G., & Sigel, H. (1999). Copper(II) complexes of synthetic fragments of amyloid beta-peptides and their complex stability. *Inorganic Chemistry, 38*(24), 5555-5563. https://doi.org/10.1021/ic990502f

20. Ritchie, C. W., Bush, A. I., Mackinnon, A., Macfarlane, S., Mastwyk, M., MacGregor, L., Kiers, L., Cherny, R., Li, Q. X., Xilinas, M., Ames, D., Davis, S., Beyreuther, K., Tanzi, R. E., Cappai, R., & Masters, C. L. (2003). Metal-protein attenuation with iodochlorhydroxyquin (clioquinol) targeting Aβ amyloid deposition and toxicity in Alzheimer disease: A pilot phase 2 clinical trial. *Archives of Neurology, 60*(12), 1685-1691. https://doi.org/10.1001/archneur.60.12.1685

21. Singh, I., Sagare, A. P., & Coma, M. (2013). Reversal of amyloid-induced cognitive deficits by targeting piRNAs and the metal-coupling biochemistry of Alzheimer's disease. *PLoS ONE, 8*(9), e74182. https://doi.org/10.1371/journal.pone.0074182

22. Tønjum, T., & Sekyere, E. O. (2022). Metals in Alzheimer's disease. *Metal Ions in Life Sciences, 22*, 337-370. https://doi.org/10.1515/9783110701375-010

23. Wang, J., & Zhang, H. Y. (2020). Role of metal dysregulation in Alzheimer's disease: Focus on the biochemistry and pharmacology of neurodegeneration. *Frontiers in Molecular Neuroscience, 13*, 31. https://doi.org/10.3389/fnmol.2020.00031

24. Wang, P., & Wang, T. (2017). Zinc dyshomeostasis and amyloid pathology in Alzheimer's disease. *Frontiers in Aging Neuroscience, 9*, 258. https://doi.org/10.3389/fnagi.2017.00258

25. Wang, Z. X., Tan, L., Wang, H. F., Ma, J., Liu, J., & Tan, M. S. (2015). Amyloid-beta: A potential trigger for tau pathology in Alzheimer's disease. *Frontiers in Aging Neuroscience, 7*, 188. https://doi.org/10.3389/fnagi.2015.00188

26. Watt, N. T., & Hooper, N. M. (2003). The prion protein and metal binding: Physiological and pathological implications. *Brain Research Bulletin, 61*(3), 303-319. https://doi.org/10.1016/S0361-9230

27.

28.

29.

30.

Buy your books fast and straightforward online - at one of world's fastest growing online book stores! Environmentally sound due to Print-on-Demand technologies.

Buy your books online at
www.morebooks.shop

Kaufen Sie Ihre Bücher schnell und unkompliziert online – auf einer der am schnellsten wachsenden Buchhandelsplattformen weltweit! Dank Print-On-Demand umwelt- und ressourcenschonend produziert.

Bücher schneller online kaufen
www.morebooks.shop

Printed by Books on Demand GmbH, Norderstedt / Germany